PREHISTORIC PALS
DINOSAUR
FACTS FOR KIDS

Words & Pictures by

Matt Hazard

Tall Tale

Tall Tale Digital, Inc. — Oakland, CA

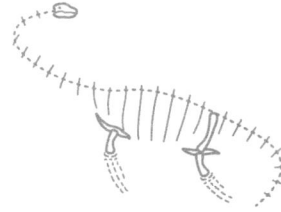

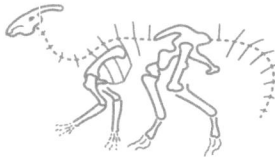

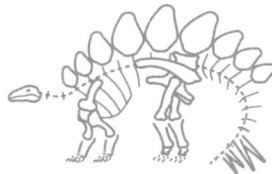

Tall Tale

Tall Tale Digital, Inc. — Oakland, CA

Prehistoric Pals Dinosaur Facts For Kids

ISBN:
978-1-955460-06-4

Library of Congress Control Number:
2024922471

This book belongs to

....

DINOSAURS

were reptiles that lived between 252 to 66 million years ago in the:

Mesozoic Era

Triassic	Jurassic	Cretaceous		Present
Period	**Period**	**Period**		**Day**

252 MYA   200 MYA   145 MYA   66 MYA   0

Millions of Years Ago (MYA)

Today, we can see the rock-like leftovers of dinosaur bones called **fossils** in a museum.

The two main groups of dinosaurs were plant-eaters called **herbivores** 🍃

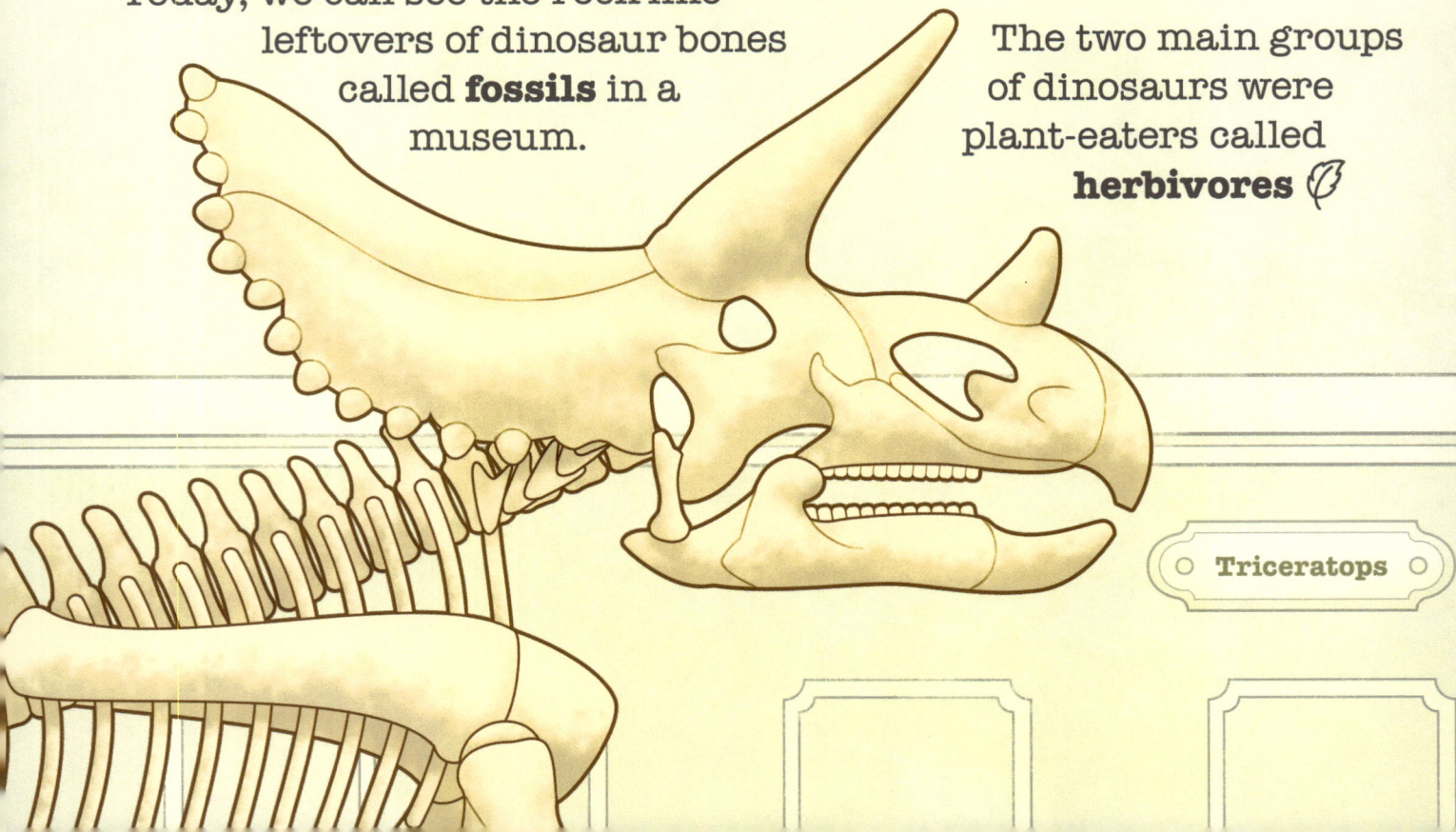

Triceratops

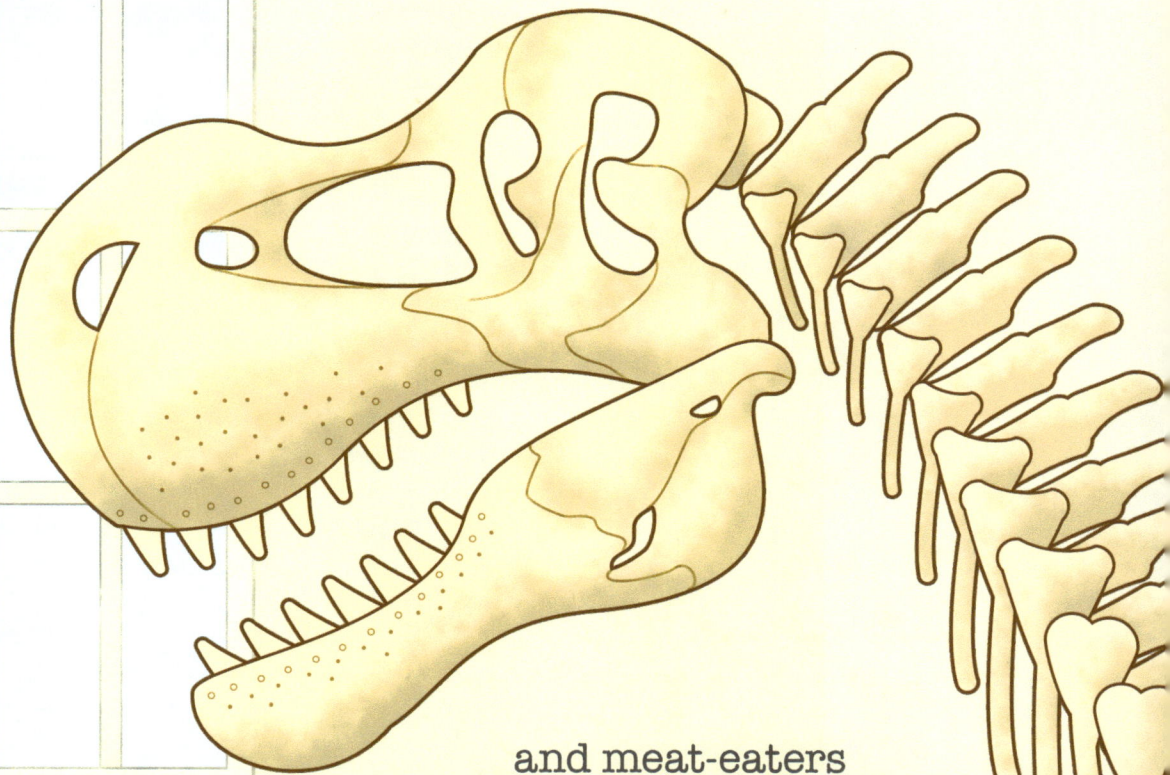

and meat-eaters
called **carnivores**.

Tyrannosaurus Rex

Hi, I'm Paulo, a third-grader who loves dinosaurs. I'm a junior **paleontologist** at the natural history museum.

Paleontologists study prehistoric plants and animals. They discover dinosaurs by digging up **fossils** from the ground.

Let's tour the museum together to learn more about some of my dinosaur pals!

Stella, the
STEGOSAURUS

STEG-oh-SORE-us
means "roof lizard"

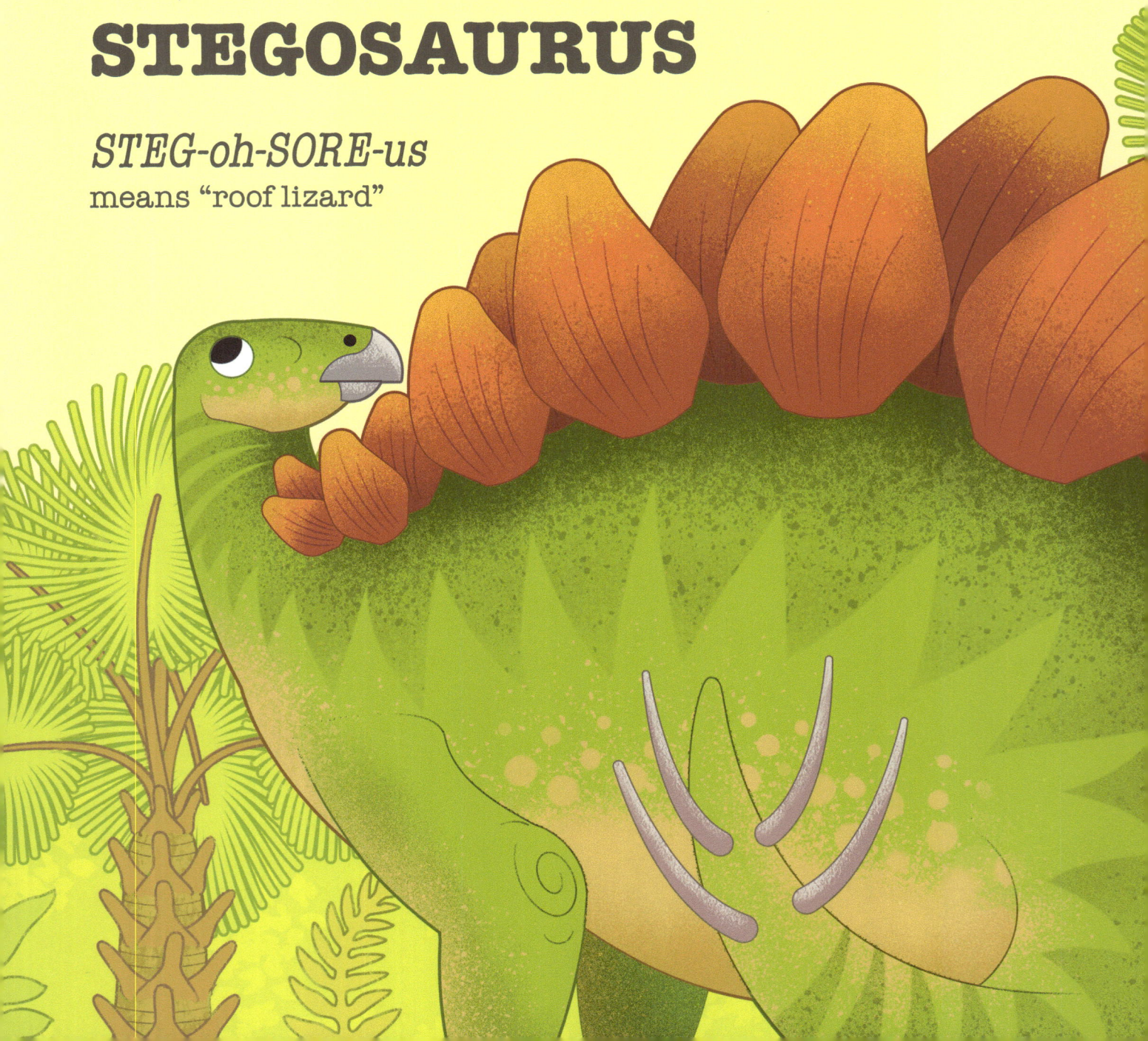

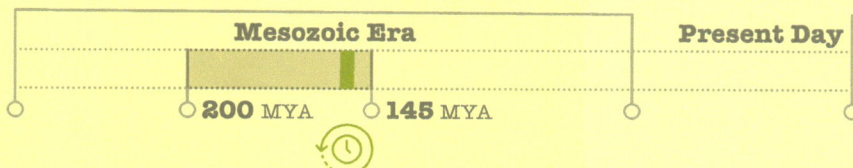

Stella lived in the **Jurassic Period**.

Stella was an **herbivore** 🍃 with 3-foot-tall backplates that might have been for keeping her body temperature just right.

Stegosaurus was up to **30 feet** 📏 long from nose to tail and weighed up to **7 tons**. 🏋️

Stegosaurus Fossils

Size of a grownup

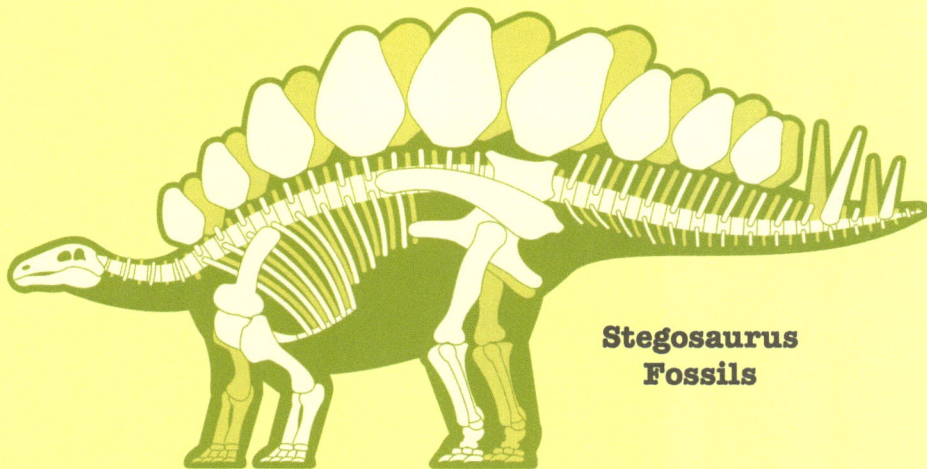

Brady, the
BRACHIOSAURUS

BRAK-ee-oh-SORE-us
means "arm lizard"

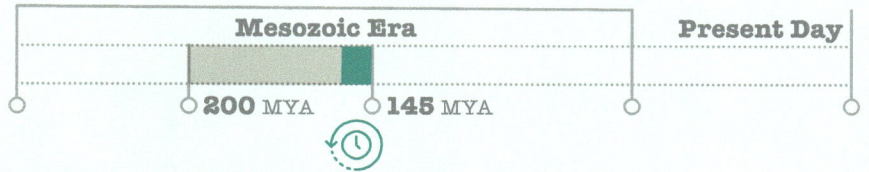

Brady lived in the **Jurassic Period**.

Brady was a giant **herbivore** that used a very long neck to reach leaves from tall trees.

Brachiosaurus was up to **49 feet** tall, **85 feet** long from nose to tail, and weighed up to **55 tons**.

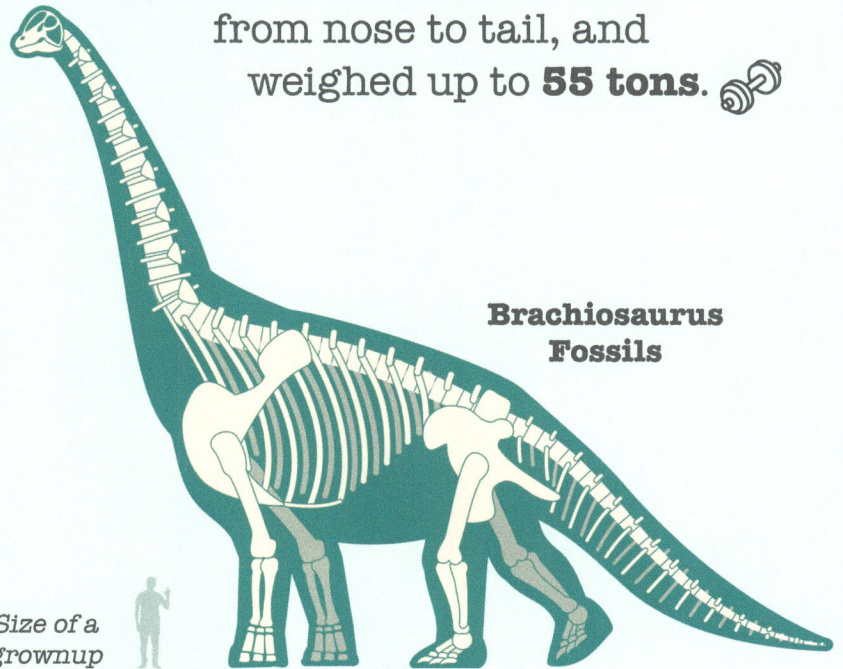

Brachiosaurus Fossils

Size of a grownup

Spencer, the
SPINOSAURUS

SPINE-oh-SORE-us
means "spine lizard"

145 MYA   66 MYA

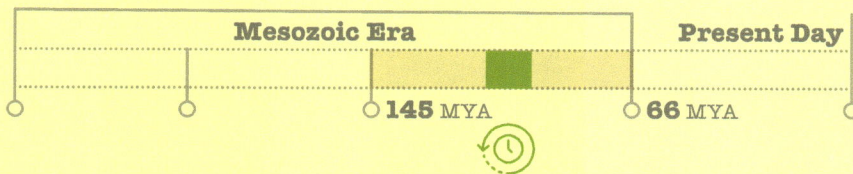

Spencer lived in the **Cretaceous Period**.

Spencer swam like a crocodile and might have been the biggest **carnivore**.

Spinosaurus was up to **59 feet** long from nose to tail with a 6-foot-tall dorsal sail, and weighed up to **10 tons**.

Spinosaurus Fossils

Size of a grownup

Teresa, the

PTERANODON

TERR-ran-OH-don
means "winged-toothless lizard"

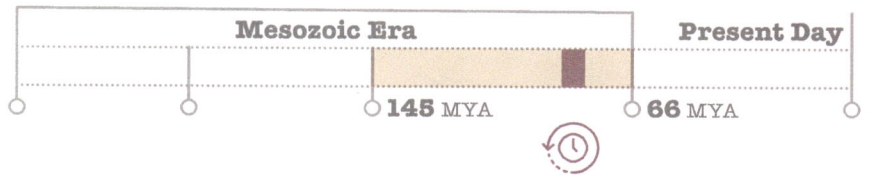

Teresa lived in the **Cretaceous Period**.

Teresa was a
carnivore with thin, hollow
bones and could fly like a bird.

Pteranodon had a wingspan of up to
24 feet wide and weighed up to **55 pounds**.

Pteranodon Fossils

Size of a grownup

Anna, the
ANKYLOSAURUS

an-KIE-loh-SORE-us
means "fused lizard"

145 MYA   **66** MYA

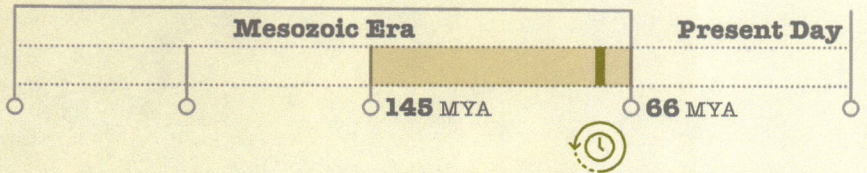

Anna lived in the **Cretaceous Period**.

Anna was an **herbivore** 🍃
that used a club-like tail
and spiky shell for protection
from carnivores.

Ankylosaurus was up to **30 feet** 📏
long from nose to tail and weighed
up to **5 tons**. 🏋️

**Ankylosaurus
Fossils**

*Size of a
grownup*

Eliza, the
ELASMOSAURUS

eh-LAZZ-mo-SORE-us
means "thin-plated lizard"

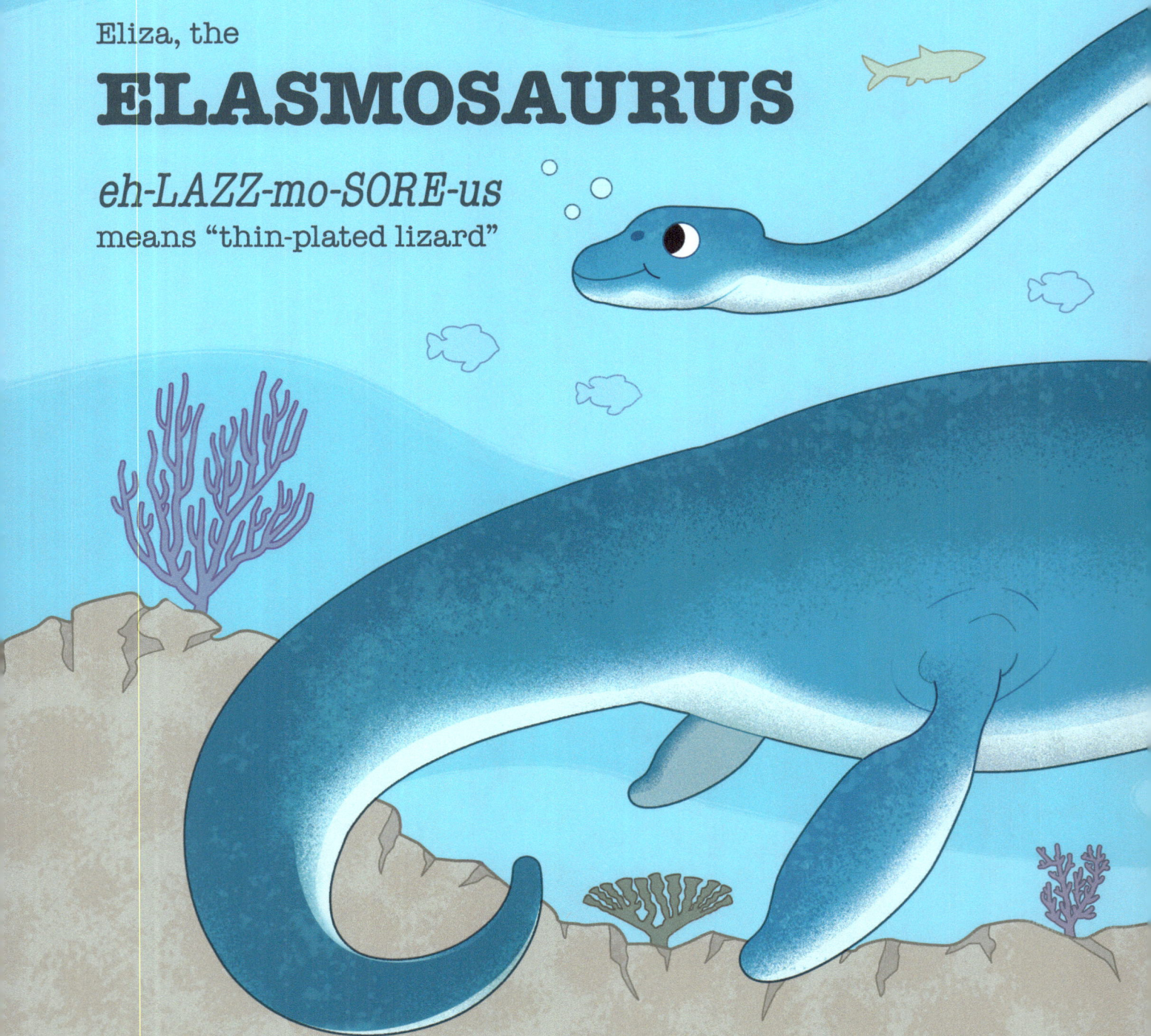

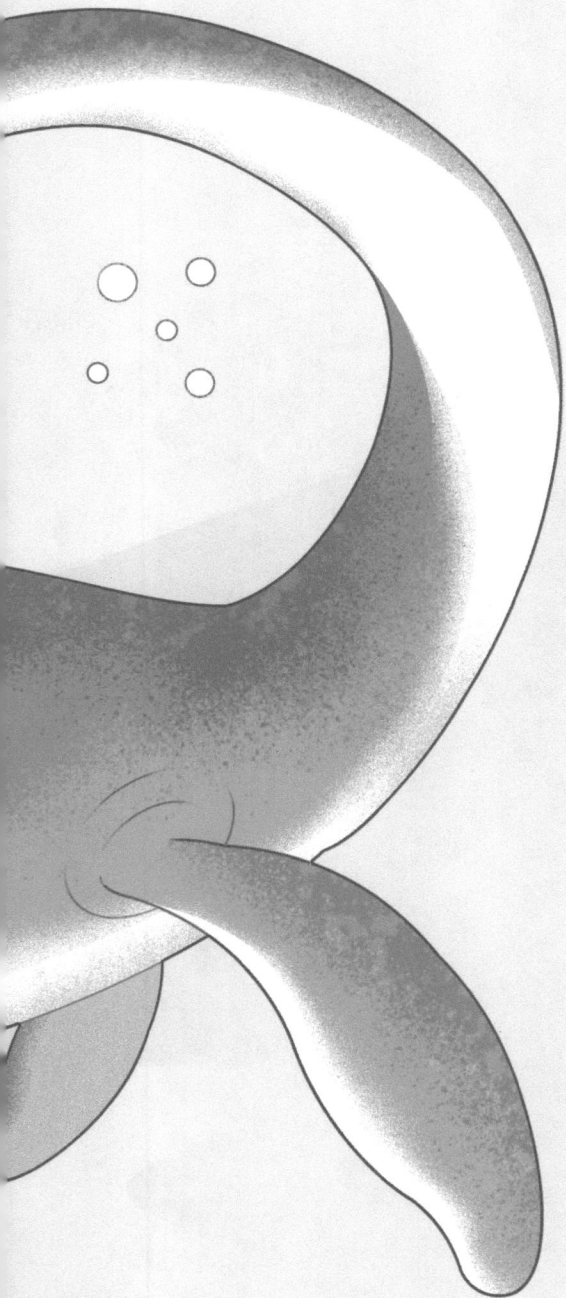

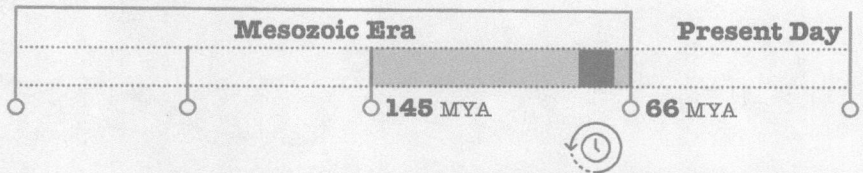

Eliza lived in the **Cretaceous Period**.

Eliza was a **carnivore**
that lived in the ocean,
feeding on smaller
dinosaurs and fish.

Elasmosaurus was up to
46 feet long from nose
to tail and weighed
up to **10 tons**.

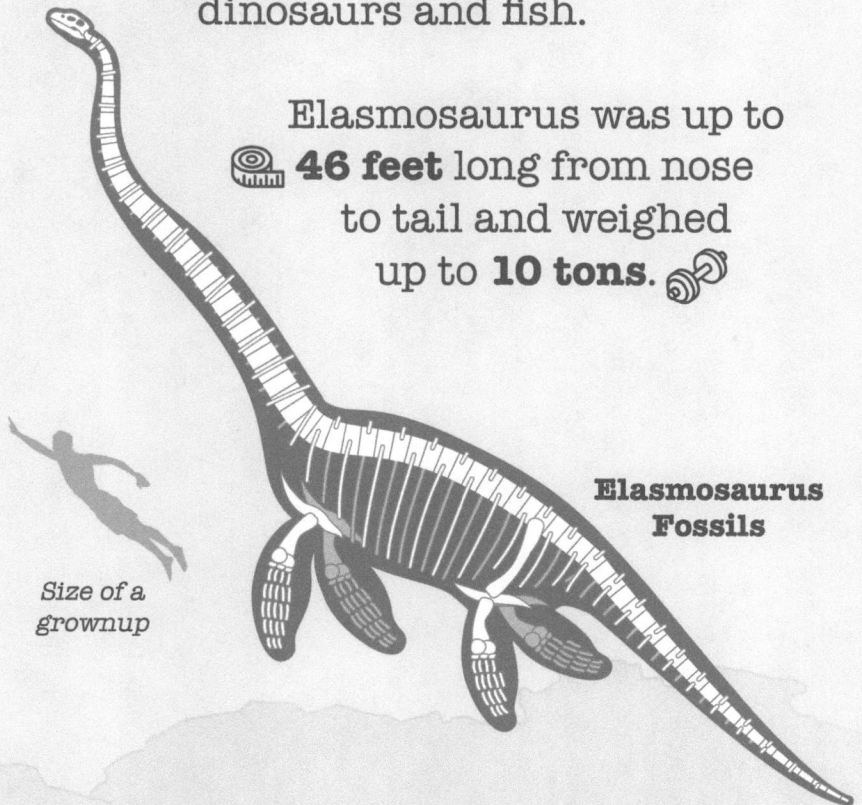

**Elasmosaurus
Fossils**

*Size of a
grownup*

Parker, the
PARASAUROLOPHUS

pa-ra-saw-ROL-off-us
means "near-crested lizard"

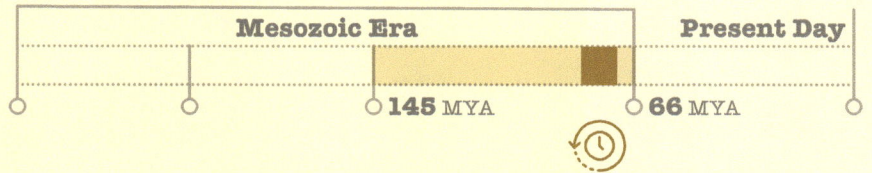

Parker lived in the **Cretaceous Period**.

Parker was an **herbivore** 🌿 called a duck-bill with a long bony crest that might have sounded like a foghorn.

Parasaurolophus was up to **36 feet** 📏 long from nose to tail, with a 5-foot crested head, and weighed up to **4 tons**. 🏋️

Parasaurolophus Fossils

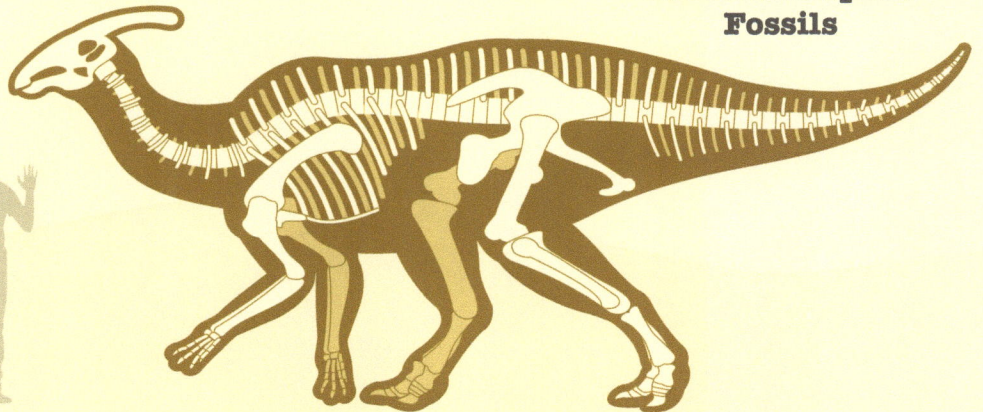

Size of a grownup

Vern, the
VELOCIRAPTOR

vel-OSS-ee-rap-tor
means "quick thief"

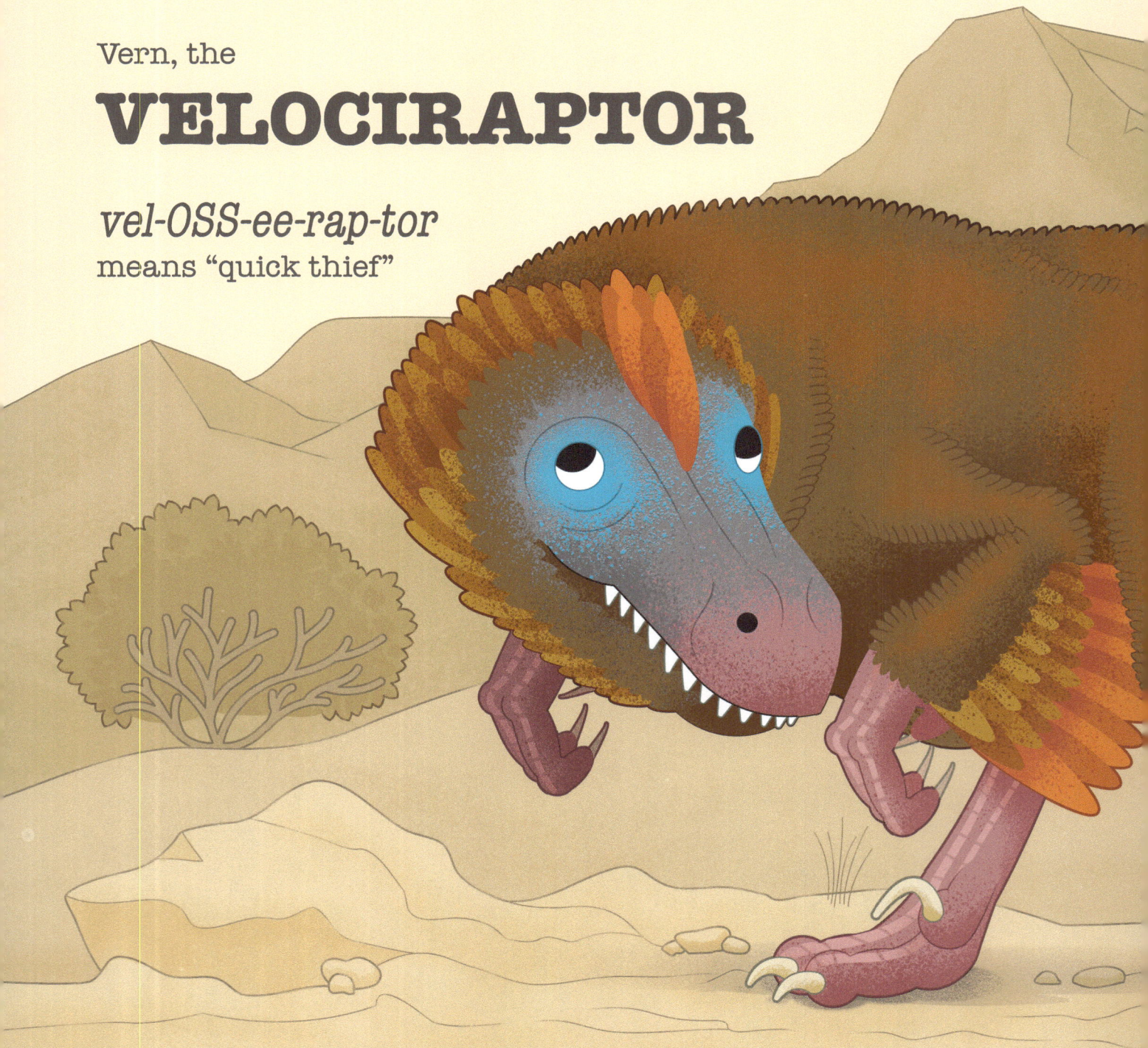

Vern lived in the **Cretaceous Period**.

Vern was a turkey-sized
🍗 **carnivore** that may have
been covered in feathers
and hunted in a pack
like a wolf.

Velociraptor was up to **6.5 feet** 📏
long from nose to tail and weighed
up to **100 pounds**. 🏋️

**Velociraptor
Fossils**

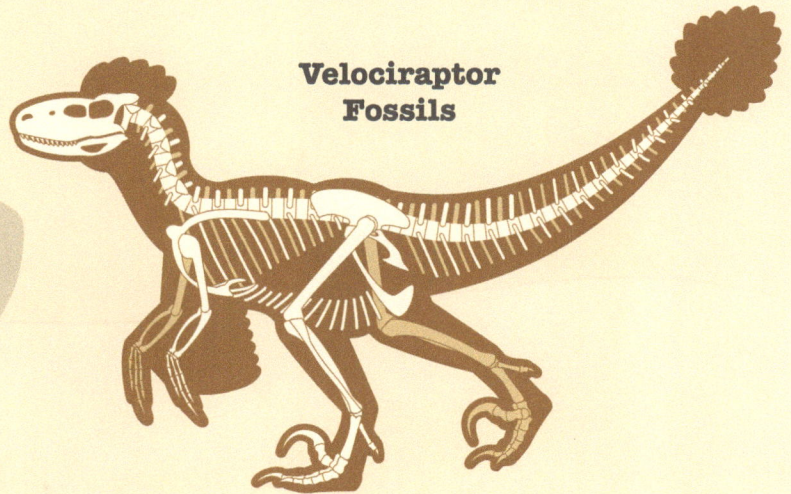

Size of a
grownup

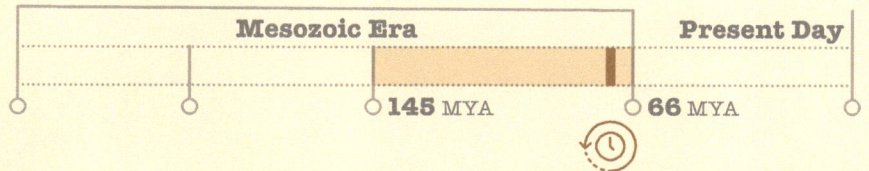

Tristan, the
TRICERATOPS

tri-SERRA-tops
means "three-horned face"

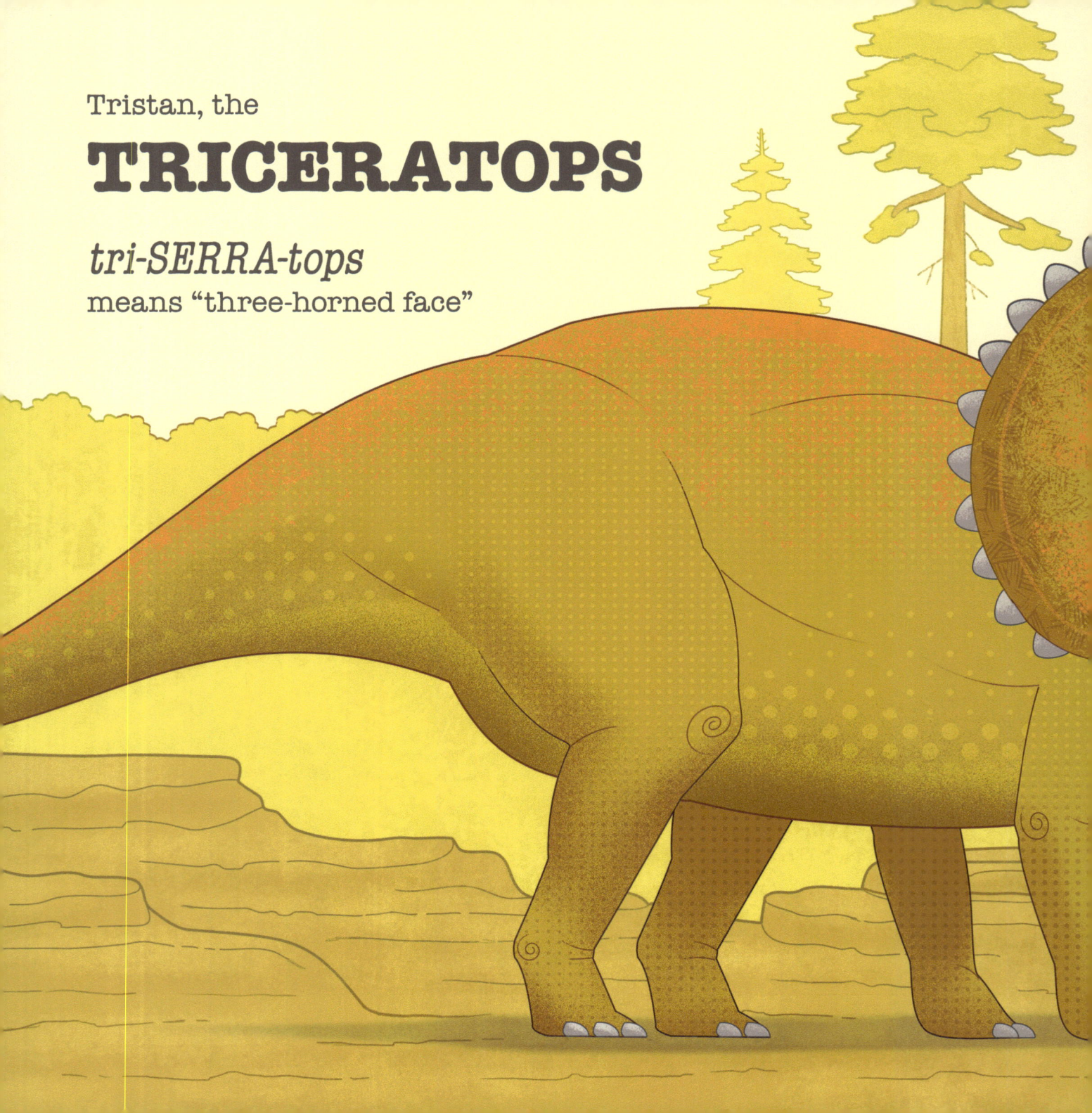

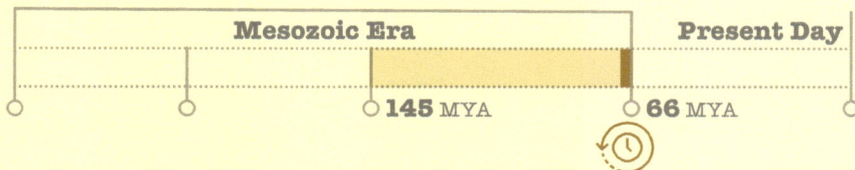

Tristan lived in the **Cretaceous Period**.

Tristan was an **herbivore** 🍃 with the biggest dinosaur head and a mouth with up to 800 teeth!

Triceratops was up to **30 feet** 📏 long from nose to tail with an 8-foot skull, and weighed up to **8 tons**. 🏋️

Size of a grownup

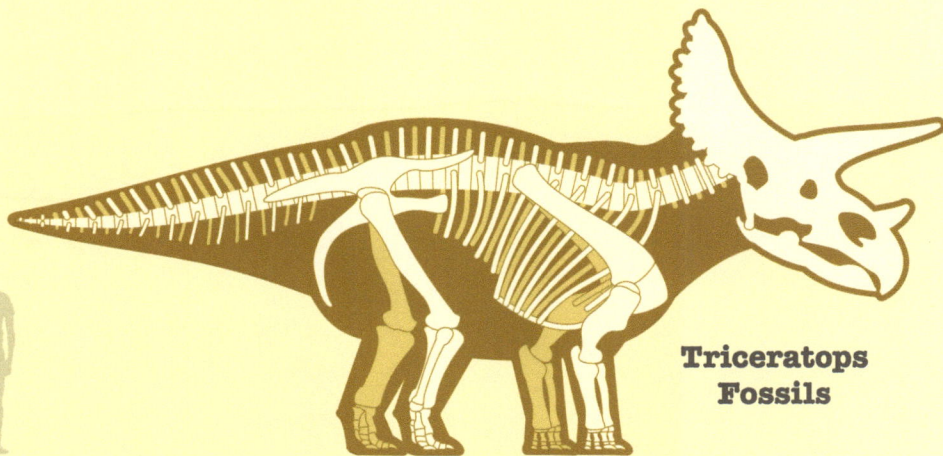

Triceratops Fossils

Tyra, the
TYRANNOSAURUS REX

tie-RAN-oh-SORE-us rex
means "tyrant lizard king"

145 MYA   66 MYA

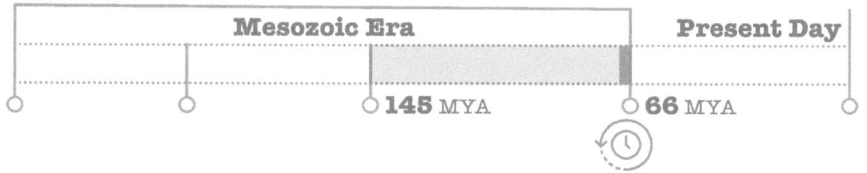

Tyra lived in the **Cretaceous Period**.

Tyra was a **carnivore**
that had teeth the size of
bananas and could eat 200
pounds in one bite!

Tyrannosaurus Rex was up to **40 feet**
long from nose to tail, up to **20 feet**
tall, and weighed up to **8 tons**.

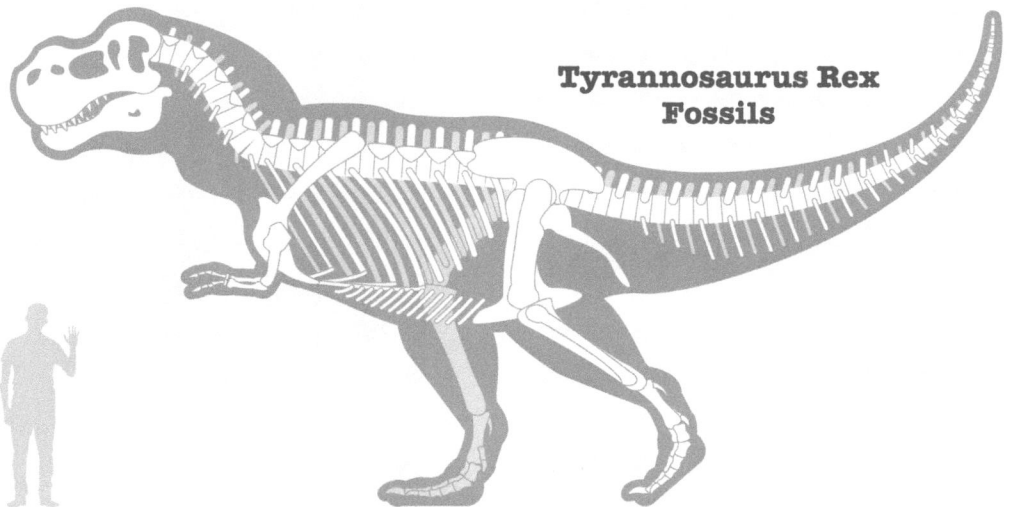

**Tyrannosaurus Rex
Fossils**

*Size of a
grownup*

What did you learn?

- Which scientists study prehistoric plants and animals?

- What are dinosaur bones called?

- What kind of food did herbivores eat?

- What kind of food did carnivores eat?

- What were the three parts of the Mesozoic Era?

Size of a grownup

Thank you so much. ♡

I appreciate you purchasing this book. I hope your kids have enjoyed learning about dinosaurs!

If you could take a few moments, I'd appreciate your honest feedback on Amazon. Your feedback helps other families find this book and is a huge help to an indie publisher like me.

To leave your feedback:

1. Open the Camera app on your phone.
2. Hold your phone so that the QR code appears in view.
3. Tap the notification to open the link associated with the QR code.

Thanks again for your support!

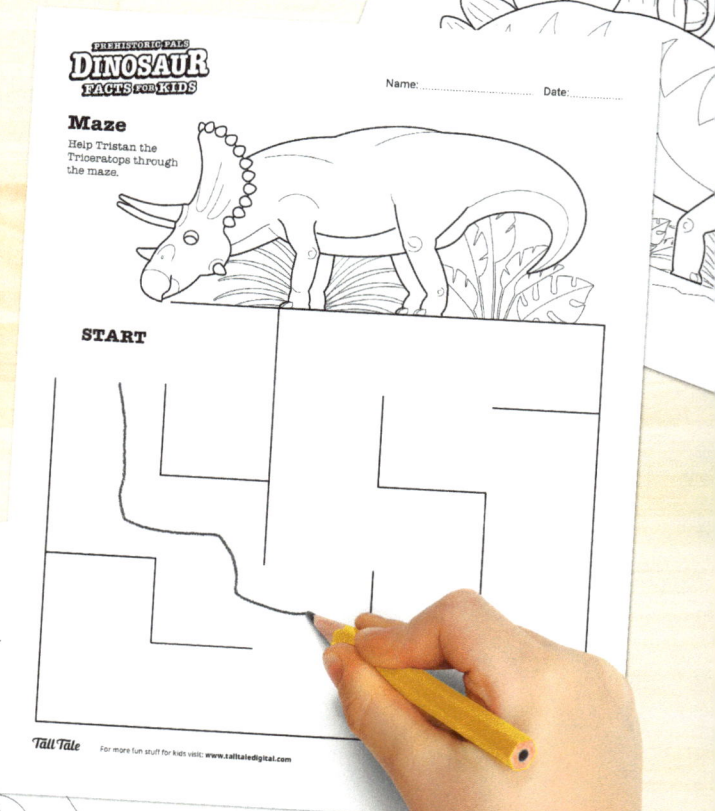

PREHISTORIC PALS
DINOSAUR
FACTS FOR KIDS

Find the Hidden
Circle these things that don't belo...

Name: _____ Date: _____

Maze
Help Tristan the Triceratops through the maze.

START

PREHISTORIC PALS
DINOSAUR
FACTS FOR KIDS

Connect the Dots
Start at 1 and connect the dots to finish this picture of Tyra the Tyrannosaurus.

Tall Tale   For more fun stuff for kids visit: www.talltaledigital.com

Get free printable activities, watch animated videos, and discover other books and apps, visit **talltaledigital.com** using this QR code:

Name:

Date:

PREHISTORIC PALS
DINOSAUR
FACTS FOR KIDS

Mask

Vern the **Velociraptor**

1. For the best results print on cardstock.
2. Cut along the dotted lines.
3. Assemble the pieces as instructed.
4. Attach string to the holes on the sides of the mask.
5. I'd love to see the fun you have with the masks!

#ttdmasks

Tall Tale

For more fun stuff for kids visit: www.talltaledigital.com

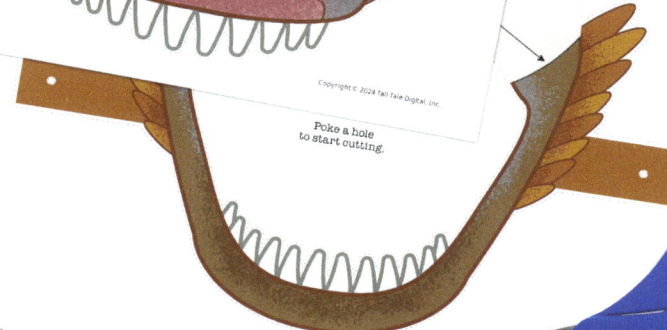

Poke a hole to start cutting.

Copyright © 2024 Tall Tale Digital, Inc.

Tall Tale

For more fun stuff for kids visit: www.talltaledigital.com

Copyright © 2024 Tall Tale Digital, Inc.

About the Author/Illustrator

Matt Hazard has been designing user experiences, making animated videos, and illustrating for over twenty years. He makes activities for kids that reinforce educational concepts, creative expression, and social norms such as sharing, empathy, and cooperation. He writes and illustrates material for his books, apps, and other activities in his studio in Oakland, CA, where he lives with his wife, daughter, and golden retriever.

www.ingramcontent.com/pod-product-compliance
Lightning Source LLC
Chambersburg PA
CBHW040812300326
41914CB00065B/1499